WHAT WOULD HAPPEN IF...
ALL THE BEES DIED OUT?

Written by Izzi Howell

Illustrated by Paula Bossio

WORLD BOOK

www.worldbook.com

READING TIPS

This book asks readers to ponder the question *what would happen if all the bees died out?* Readers will discover why bees are in danger and will contemplate a world without them. Use these tips to help readers consider the ripple effects of certain actions and events.

Before Reading

Explain to readers that this book uses cause and effect to show the role bees play in our world. Cause and effect can help us think about why things are the way they are. It can also help us think about what might happen in the future because of our actions today. Encourage readers to be on the lookout for examples of a cause and effect structure as they explore what would happen if all the bees died out..

During Reading

Discuss with readers how some actions and events have multiple causes and others have multiple effects. Explain that it can be tricky to keep all the if/then scenarios straight in our minds, so it can be helpful to create a visual guide. Encourage readers to draw and add notes to their own cause and effect maps like those found on pages 20-21, 24-25, and 34-35.

After Reading

After finishing the book, discuss with readers how their understanding and opinions of bees have changed. Additionally, you can have readers respond to the comprehension questions included on page 46 and can complete the Chain of Events activity on page 47 to further extend the learning.

Visit **www.worldbook.com/resources** for additional, free educational materials.

There is a glossary of terms on pages 44–45. Terms defined in the glossary are in boldface type that **looks like this** on their first appearance on any spread (two facing pages).

Contents

Bye-bye bees? 4

Bees in crisis 6

Perfect pollinators 16

A balanced ecosystem 22

Time to act! 28

A world without bees 36

Conclusion 40

Summary 42

Glossary 44

Review and reflect 46

Bye-bye bees?

They're unwelcome guests at your picnic and can give you a nasty sting if you're not careful. A bee sting can even kill a person with a serious allergy. Would we really miss bees if they all died out?

Beatrice? Where have you gone?!

THINK ABOUT IT!

If you could click your fingers to remove one animal from our planet, what would it be, and why? What do you think would happen if this animal disappeared?

We may know the answer soon enough, since bees are in serious danger. Wild bee **populations** are decreasing every year across the world. Bees face many threats, including harmful **pesticides, habitat** loss, **climate change, parasites,** and diseases.

Bees are found on every continent on Earth except for Antarctica. They live in the countryside, on farms, in woodlands, and around our homes in parks and cities. You've probably seen bees many times before without giving them a second thought. But how much do you know about these buzzing insects?

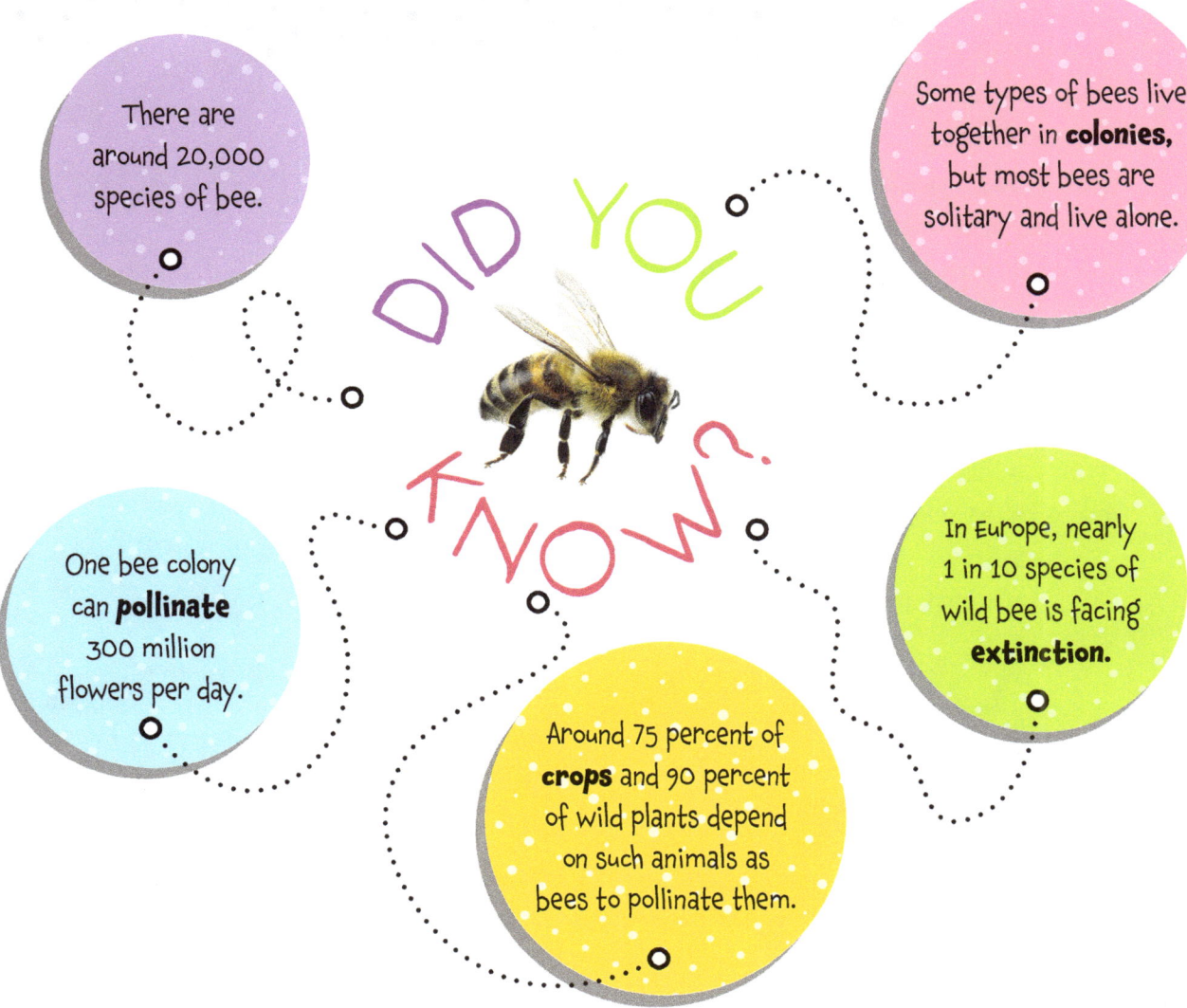

DID YOU KNOW?

There are around 20,000 species of bee.

Some types of bees live together in **colonies,** but most bees are solitary and live alone.

One bee colony can **pollinate** 300 million flowers per day.

Around 75 percent of **crops** and 90 percent of wild plants depend on such animals as bees to pollinate them.

In Europe, nearly 1 in 10 species of wild bee is facing **extinction.**

Were you surprised by any of the facts? It turns out that bees are a big deal! These tiny insects are one of the most valuable animals on Earth to humans, plants, and animals alike. Without bees, our food supply and all other natural **ecosystems** would collapse. Want to find out how and why? Read on to find out!

Bees in crisis

So why exactly are bees in danger? In this chapter, we'll explore some of the main threats facing bees, including **habitat** loss, **pesticides,** disease, **climate change,** and **invasive species.** Let's start with a major one—habitat loss.

FUN FACT!
One healthy meadow can be home to up to 100 different species of wildflowers!

Over the past few hundred years, many wild habitats around the world have been destroyed. They have been cleared to build new towns and roads or turned into farmland.

Many of these spaces, such as meadows, forests, and fields, were important bee habitats. These habitats had plenty of wild plants that provided **nectar** and **pollen** for hungry bees, and places where bees could nest safely. Now that they are gone, bees are struggling to find food and shelter.

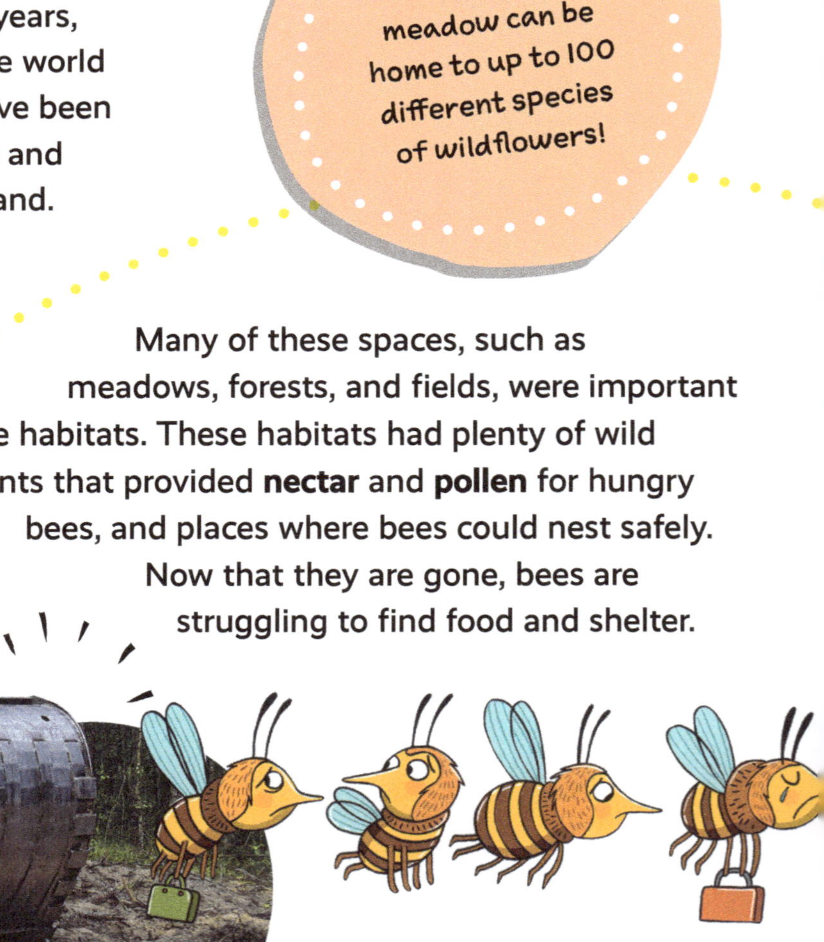

Even though many farms grow **crops** that could provide bees with nectar and pollen, these plants alone aren't enough for bees. Just like us, bees need to gather nectar and pollen from lots of different plants to have a healthy, balanced diet. This was much easier when they had access to wild spaces with many different plants and flowers.

I'm so bored of rapeseed!

This is a field of rapeseed flowers. Rapeseed is grown on farms and used to produce oil.

DID YOU KNOW?

Since 1990, more than one billion acres of forest have been destroyed.

THINK ABOUT IT!

What do you think we could do to solve the problem of habitat loss for bees? Think about it on your own, and then turn to pages 28-29 for some more ideas.

BEES IN CRISIS

Most farmers around the world use **pesticides** to control harmful insects that damage their **crops.** These pesticides help farmers grow more food. However, pesticides also have a devastating effect on bees and other insects that visit farms.

Potato farmers use pesticides to control the potato beetle, which can do huge amounts of damage to potato crops. Without pesticides, these beetles could destroy whole fields of potatoes.

Pesticides are even found in untreated areas. They are left behind in the soil and dust, which can be carried by the wind to pesticide-free areas. So, there's nowhere for bees to escape!

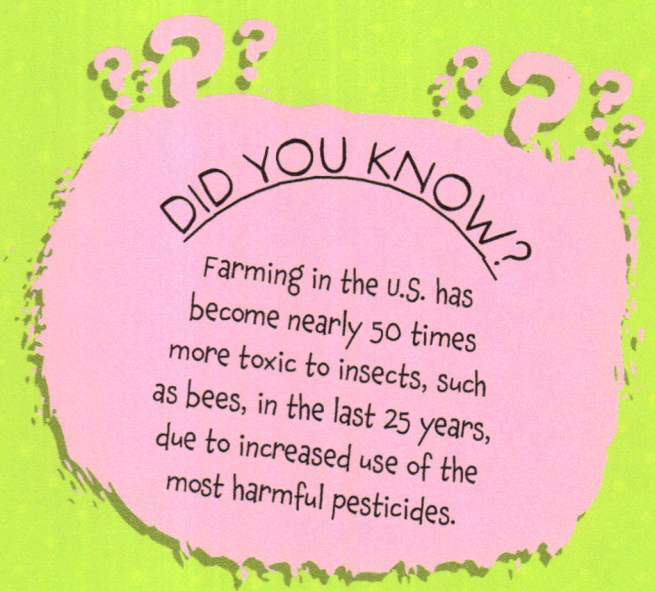

DID YOU KNOW?

Farming in the U.S. has become nearly 50 times more toxic to insects, such as bees, in the last 25 years, due to increased use of the most harmful pesticides.

Scientists have discovered that pesticides have many different harmful effects on bees. After being exposed to pesticides, bees can lose the ability to find food or find their way home. Some bees can't eat or fly properly. Others can't **reproduce,** or have young. If bees can't find food, eat, or reproduce, they are in serious trouble!

If a bee picks up pesticide on its body, it can carry it back inside the hive and pass it on to bee **larvae** before they turn into adult bees.

Bee larva

Hi! I'm David Goulson. I'm a British scientist who studies bee ecology. Ecology is the study of all the living things in an environment, plus the air, water, and soil, too. As part of my research, I can see how **habitat** loss and pesticides have affected bees. I think it's really important to share my research with everyone, not just other scientists. I write popular science books about the dangers facing bees and set up **citizen science** projects, such as counting the number of bees in yards and gardens. The more people that know about the risks facing bees, the more we can do to save them!

THINK ABOUT IT

What are the advantages and disadvantages of pesticides? Do you think the risks are greater than the reward?

BEES IN CRISIS

Just like us, bees can get sick! They are affected by many different diseases and **parasites.** Some of these conditions are very serious and are currently killing large numbers of bees.

At the moment, scientists know more about honey bee diseases than wild bee diseases, because it's much easier for them to study honey bee hives looked after by beekeepers. However, we do know that wild bees are already affected by certain diseases and parasites, and they may also be at risk from honey bee diseases.

No mites, yay!

The varroa mite is one of the greatest threats to honey bees, even though it's only the size of a sesame seed. This tiny parasite drinks bees' blood, which is unpleasant, but not actually the main problem. The danger is that varroa mites feed from many different bees and help spread potentially deadly **viruses** among them.

DID YOU KNOW?

In some areas of Asia, Europe, and North and South America, up to half of all honey bee hives are infested with varroa mites.

These viruses are also found in **colonies** with no mites. However, being bitten by mites can weaken a bee's **immune system,** which makes it harder for the bee to fight the virus and recover from the illness. If a honey bee hive is infested with varroa mites, most of its bees will probably die.

Hives affected by foulbrood must be burned once they are empty to keep the disease from spreading.

Honey bees are also affected by such diseases as American foulbrood and European foulbrood, which are caused by bacteria. These diseases turn bee **larvae** into brown, smelly goo. Yuck!

THINK ABOUT IT!

Beekeepers play an important role in treating and controlling honey bee diseases. Why is it harder to treat wild bees and keep diseases from spreading among them?

BEES IN CRISIS

Global warming is making spring arrive earlier in the year than ever before. This isn't just affecting the weather. It's affecting all living things ... and yes, of course, bees, too!

In temperate areas with different seasons, many living things behave differently as the seasons change. Plants and many plant-eating animals, including bees, have evolved so that they are in sync throughout the year and can make the most of the food available.

Bees are most active during spring and summer when plants produce tons of **nectar** and **pollen**. During the winter, there is far less food around, so bees are less active. Honey bees usually shelter inside their hive, while queen bumblebees **hibernate** in burrows in the ground. Some adult wild bees die in the fall and winter, leaving their eggs to hatch in the spring.

Up until recently, bees have always found flowering plants ready to snack on when they emerged in spring. However, plants are now flowering earlier because of **climate change**. When plants flower before the bees are ready, bees miss them entirely. This leaves the bees very hungry *and* makes it hard for the plants to be **pollinated** and **reproduce** (see pages 16–21).

Zzzzz zzzz.

A hibernating bumblebee

Why didn't anyone wake me up sooner?!

THINK ABOUT IT!

Scientists have also noticed that unusually warm winter weather can confuse bees and make them come out long before any plants have flowered. Why is this a problem?

DID YOU KNOW?

In the UK, plants are now flowering a month earlier than they did in the 1980's.

BEES IN CRISIS

There are also many other specific threats to bees in certain areas, such as **invasive species.** In around 2004, Asian hornets were accidentally brought from Southeast Asia to Europe. Since then, they have started to spread across parts of the continent. The problem is that Asian hornets *love* to feast on bees. In Southeast Asia, bees have learned how to avoid or attack Asian hornets when necessary, but European bees don't know how, making them an easy target. If Asian hornet **populations** rise in Europe, it'll be another big problem for the bees that live there.

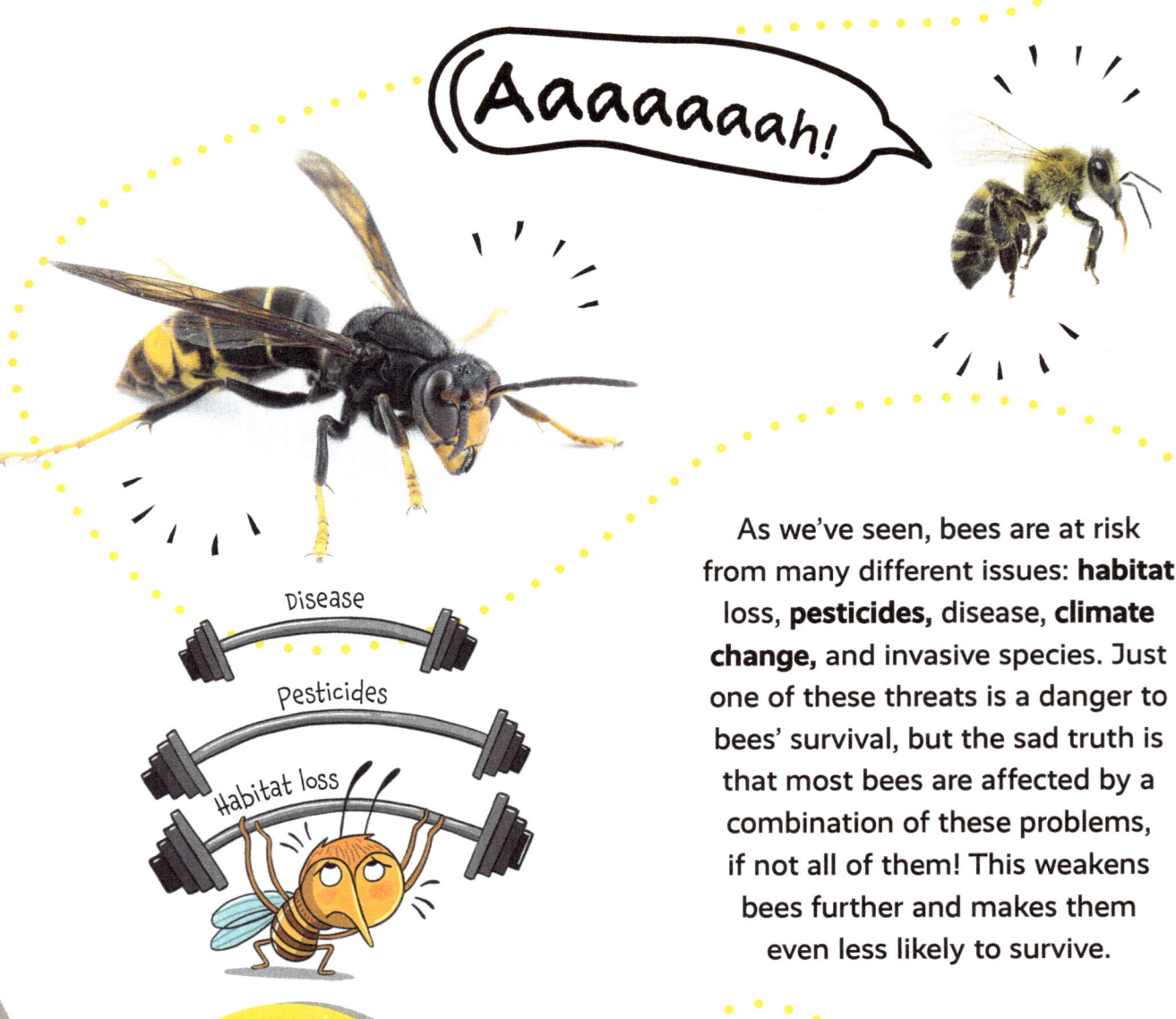

As we've seen, bees are at risk from many different issues: **habitat loss, pesticides,** disease, **climate change,** and invasive species. Just one of these threats is a danger to bees' survival, but the sad truth is that most bees are affected by a combination of these problems, if not all of them! This weakens bees further and makes them even less likely to survive.

If these threats continue, it's unlikely that bees will suddenly disappear overnight. However, unless we act soon, it's very possible that rarer species of wild bees will start to die out. If the issues worsen, even such common species as honey bees may not survive. The only surviving bees will be in captivity, making the species technically **extinct.** The loss of bees from our planet will have devastating consequences on the natural world. Turn the page to find out why!

DID YOU KNOW?

More than 30 species of bees are already considered **endangered** or critically endangered.

The rusty patched bumblebee was declared endangered in 2017. Since the 1990's, its population has dropped by 90 percent.

Perfect pollinators

Bees perform one of the most important roles in the animal kingdom—helping flowering plants to **reproduce.** Without bees, flowers wouldn't be able to produce seeds or new plants.

Bees are the perfect size to squeeze inside flowers and collect the food inside. They gather tiny grains of **pollen** and a sweet liquid called **nectar.** Bees make honey from the nectar, and they use both honey and pollen as food. While bees are slurping up nectar from deep inside the flower, pollen gets stuck to their body. This pollen is carried with them from flower to flower.

THINK ABOUT IT!

Bees are the only animal that can make honey. Do you like to eat honey? What could you use instead if there were no more bees to make honey?

FUN FACT!

Bees get energy from nectar and protein and other **nutrients** from pollen.

Pollen delivery arriving!

As well as being a tasty snack for bees, pollen is also needed for plant reproduction. Pollen is produced in the male plant of a flower. When pollen-covered bees visit other flowers, they transfer the pollen to the female parts of those flowers. This process is called **pollination** and needs to happen for flowers to produce seeds.

DID YOU KNOW?
Bees are also very speedy pollinators. One bee can pollinate around 5,000 flowers in just one day!

Other animals, including birds, bats, and such other insects as butterflies, can also pollinate plants. However, bees are one of the most important. Some species of bees are perfectly adapted to pollinate certain plants. Garden bumblebees have long tongues to reach inside deep flowers.

Slurp, slurp!

Bees and other pollinators don't just **pollinate** wild plants. They are also responsible for pollinating many **crops** that we eat every day!

These are just some of the crops pollinated by bees and other animals:

PERFECT POLLINATORS

Crops pollinated by bees provide us with many important **nutrients.** We get vitamins and minerals from fruits and vegetables, such as strawberries and cabbages; fats and protein from nuts and seeds, such as almonds and sunflower seeds; and carbohydrates from potatoes. We need to eat all these things as part of a balanced diet to stay healthy and grow strong!

DID YOU KNOW?

One in three mouthfuls of the food we eat depends on pollination from such insects as bees!

Hi! I'm Jiandong An. I'm a Chinese scientist with a particular interest in bee pollination. Bees are so important to both wild plants and crops. As well as studying native bees of China and their role in our **ecosystems,** I also research the importance of bees to **agriculture.** Bees are very quick and effective pollinators, which means more crops for farmers! I have even discovered that tomatoes pollinated by bumblebees taste better than tomatoes hand-pollinated by humans!

THINK ABOUT IT!

Do you think farmers would be convinced to protect bees after reading Dr. Jiandong An's research? What are the benefits to farmers of using bees to pollinate their crops?

What would happen if bees weren't around to help plants reproduce?

Bees are essential for the **pollination** of many flowering plants, including **crops** that we depend on for food. If all the bees died out, these plants would find it much harder to **reproduce**. Let's explore why this would be bad news for living things, **habitats,** and people.

If pollination doesn't happen, a plant can't produce seeds. In the short term, this would affect animals that eat the fruit and seedpods that seeds grow in. They would go hungry and many would die. Their **population** would go down, affecting the balance of the **ecosystem** (see pages 22-27).

If flowering plants don't produce any new seeds, no new plants will grow. Just as with seeds and fruit, the sudden disappearance of these plants would affect the animals that depend on them for food and the ecosystems that the plants and animals live in. The animals that live in or on these plants would have to find new places to shelter.

Time to find a new tree!

Plants are a big deal in our ecosystems. Their branches and leaves create shade, which keeps habitats cool. Their roots hold the soil together.

THINK ABOUT IT! If there were no plants to hold the soil together, what do you think will happen?

DID YOU KNOW? Such animals as bees pollinate more than 75 percent of the main crops we depend on for food.

If crops aren't pollinated, they won't produce fruits, nuts, or seeds. We would suddenly be unable to grow many important foods, such as potatoes, apples, beans, and tomatoes. That's right—no more french fries dipped in ketchup! But more seriously, without these crops, many people would go hungry and would not be able to eat a healthy, balanced diet.

It's a fruit salad ... without the grapes, orange, melon, strawberry, or mango.

A balanced ecosystem

Bees play a key role in many different **ecosystems.** As well as helping plants to **pollinate,** they are also the **prey** of lots of hungry animals! But this isn't the reason why bees are under threat—it's an important part of the circle of life!

All the living things in an ecosystem eat (or are eaten by!) other living things. **Herbivores,** like most bees, feed on plants. **Omnivores** eat plants and other animals. **Carnivores** only eat animals. We can draw a **food web** to show what eats what. A food web shows how all the different food chains in an ecosystem are connected.

Plants are at the bottom of most food webs. They make their own food using sunlight, water, and gases in the air. Herbivores feed on plants, so they are drawn above them in the food web. Even higher up the food web are omnivores and carnivores.

Let's look at how bees fit into a food web!

Different herbivores eat different parts of the plant. Bees eat **nectar** and **pollen** from flowers, while other animals eat the leaves, stem, or roots.

Can you guess the favorite food of the European bee-eater? That's right ... bees! For this reason (and many others), bees are an important part of the grassland and forest ecosystems of Europe, Africa, and West Asia where European bee-eaters live. However, bees aren't the only thing that European bee-eaters eat. The food web shows what else they feed on and which animals they are eaten by.

FUN FACT!

European bee-eaters hit bees and wasps against a hard surface to remove their stinger. Then they throw the insects into the air and catch them with their beak!

"Eeeek! This one still has its stinger!"

A BALANCED ECOSYSTEM

We've already seen that all the living things in an **ecosystem** are connected. So, what do you think would happen if one animal—for example, bees—disappeared from that ecosystem? That's right! The rest of the ecosystem would be affected. Ecosystems are very delicately balanced. Losing just one species has a massive domino effect on all the other animals living in it.

What would happen to the food web on the previous page if bees disappeared?

Bees disappear

European bee-eaters would eat more wasps and grasshoppers to make up for the missing bees in their diet. As a result, the wasp and grasshopper **population** would also go down.

Without bees drinking **nectar**, more nectar would be available for other insects to eat. Insects that might have previously died because there wasn't enough nectar to go around would now survive. Slowly, their population would increase.

If bee-eaters died of starvation, the **predators** of bee-eaters, such as black kites, wouldn't have as much food to eat. As a result, some black kites would die of hunger, and their population would decrease. This in turn, would affect their predators higher up the food chain, such as Eurasian eagle owls.

European bee-eaters would lose a large part of their food supply. Most birds would go hungry, and many would die. Their population would decrease.

Fewer grasshoppers means less food for animals such as lizards. You've guessed it! Their population would go down as well.

THINK ABOUT IT!

If black kites disappeared instead of bees, what do you think would happen to the European bee-eater population?

DID YOU KNOW?

European bee-eaters are considered a pest by some farmers because they eat honey bees. If a hive's queen is eaten by a European bee-eater, the rest of the **colony** will collapse.

Back off, bee-eaters!

Bees, European bee-eaters, black kites, and the other animals from the **food web** on the previous page are just a few examples of animals that live together and depend on each other for food. In reality, each animal in that **ecosystem** is connected to hundreds, if not thousands, of other species in massive food webs. Every single animal would be affected by the loss of bees.

FUN FACT! Black bears don't just eat honey. They also love to feed on bees!

Bees? Where? I'm starving!

These rainforest bees have built their nest between large leaves.

What's more, bees live in many different ecosystems, including rainforests, deserts, mountains, fields, and woodlands. If bees disappeared, all the animals in these ecosystems would be affected as well.

A BALANCED ECOSYSTEM

As we saw in the food web, the **populations** of some species would go down following the loss of bees. Some animals that depend heavily on bees for food, such as bee-eaters, would probably become **endangered** or maybe even **extinct**.

Other animal populations might increase as a result. However, this isn't necessarily a good thing. After sea otters were hunted almost to extinction in the 1700's and 1800's, the number of sea urchins (their **prey**) rose dramatically, since there were barely any sea otters around to eat them! The increased numbers of sea urchins ate through huge amounts of kelp (their favorite meal!), which nearly destroyed the kelp forest ecosystem. Remember—ecosystems are carefully balanced. Just one change ripples out across all the other species that live there.

THINK ABOUT IT!

Think back to the animal you chose to remove from our planet on page 4. Now that you know a little more about food webs and ecosystems, what do you think would happen if this animal disappeared?

Time to act!

Here's the good news! It's not too late to protect bees and keep them from dying out. If we take the following steps, it's very likely that bee **populations** will recover.

Although bees do harvest **nectar** and **pollen** from **crops,** they are wild animals and need wild spaces for food and shelter. The destruction of wild **habitats,** such as fields, meadows, and forests, is one of the key reasons why bees are at risk (see pages 6–15). So, reducing development on wild sites is crucial if we don't want the problem to get worse. We can also adapt developed areas to give bees more places to live and find food.

A bee hotel

Did you know that most bees don't live in hives?! Many species of bees live alone in small burrows in the ground, gaps between stones, holes in tree trunks, or inside hollow branches. Leaving undisturbed piles of natural materials will create a nesting space for wild bees. You can also buy or make a bee hotel out of bamboo or wood.

I'm ready to check in!

TIME TO ACT!

Scientists have shown that **pesticides** are one of the greatest threats to bees (see pages 8–9). However, pesticides are still being used in many places around the world. A worldwide ban on the most dangerous pesticides would have a huge positive impact on bee **populations**.

DID YOU KNOW? Bees can develop resistance to pesticides which causes farmers to use more pesticides each year. In some cases, new pesticides need to be developed altogether.

Farming without pesticides is known as **organic** farming. Organic farmers control pests using natural methods, for example, by attracting birds, frogs, and other animals that snack on plant-munching insects, such as caterpillars and aphids. This makes them much closer to a "wild" **habitat** than a conventional farm. So, as well as not using pesticides that actively hurt bees, organic farms are also a good place for bees to find shelter.

A global change to organic farming would help protect many different species from pesticides, including bees. However, this kind of huge shift in **agriculture** wouldn't be easy. Many farmers haven't tried organic farming before, so they would need help and support. Organic farming also produces about 25 percent fewer **crops** than normal farms, since more crops are lost to pests. So, we'd need more farms and farmers to grow the same amount of food.

THINK ABOUT IT!

Ladybugs love to munch on aphids, which are a common pest killed by pesticides. What do you think would happen to ladybug populations if farms stopped using pesticides?

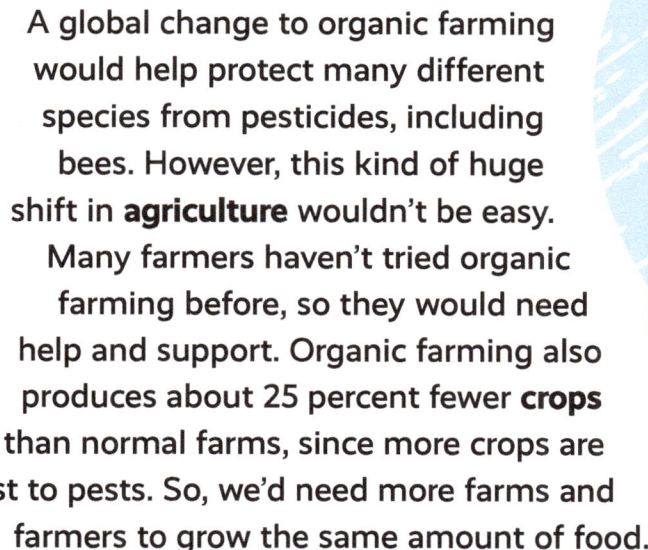

Hey neighbor!

FUN FACT!
There are up to 75 percent more wild bees on organic farms!

TIME TO ACT!

Scientists are working on new ways to protect bees from diseases and **parasites.** One promising development was the approval of the first **vaccine** for bees in 2023. The vaccine protects them against American foulbrood (see page 11).

To create the vaccine, scientists put a dead version of the bacteria that causes American foulbrood into royal jelly, a special food only eaten by the queen bee. This makes the queen develop her own antibodies, which fight the disease. She then passes on her antibodies to the eggs that she lays. In this way, the entire hive is protected against foulbrood and won't become sick if exposed to it later.

Antibodies for you, and you, and you!

Scientists test new treatments before they can be used by beekeepers.

Scientists hope that the technology used to create the American foulbrood vaccine could be used to develop vaccines for other deadly bee diseases. They are also working on other new ways to protect bees, such as breeding bees that are more resistant to diseases and parasites.

FUN FACT!

There is a type of bee named after Marla Spivak! Its official name is *Lasioglossum spivakae*.

Hi! I'm Marla Spivak. I'm an American scientist who specializes in honey bee research. One of my main focuses is the health of honey bees. In my bee lab, we have been studying the genetics of honey bees to find ones that can naturally defend themselves against illnesses and mites. We have used this information to breed stronger bees that won't get sick and don't require any medicine. We hope that beekeepers will start to use our honey bees in their hives. More healthy bees in hives means more **pollinated** flowers!

DID YOU KNOW?

Peppermint and cinnamon oils can also be used to protect honey bees against varroa mites!

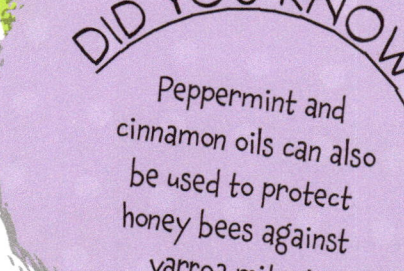

TIME TO ACT!

Working to solve the global climate crisis is hugely important for the survival of bees and the rest of the planet. Unless we change our behavior, **global warming** and **climate change** will continue to get worse, which will cause serious disruption to the seasonal changes of animals and plants.

DID YOU KNOW?
Scientists believe that one in every six species is at risk of **extinction** if carbon emissions continue at the current pace.

What causes global warming?

Fossil fuels, such as coal, oil, and natural gas, are burned to generate electricity and to power vehicles and factories. This releases carbon dioxide and other greenhouse gases.

Methane, another greenhouse gas, is produced on massive cattle farms and by burning natural materials, such as wood.

Trees and plants remove carbon dioxide from the air and use it to produce energy. Because so many natural areas have been cleared and built on, there are fewer plants to remove carbon from the air, making carbon dioxide levels rise even higher.

If all countries act together and reduce their greenhouse gas emissions now, we have a hope of undoing the damage we've done to our planet over the last few hundred years. Lowering the amount of greenhouse gases in the atmosphere should help our seasons and weather return to normal, leaving bees and other animals much less confused and hungry!

THINK ABOUT IT!

What do you think would happen to the levels of carbon dioxide in the atmosphere if we planted more trees?

Finally, a flower at the right time!

Greenhouse gases gather in our atmosphere and trap heat from sunlight close to Earth's surface. This is known as the greenhouse effect.

The greenhouse effect is responsible for a global rise in temperatures, also known as global warming. This is linked to climate change, which includes changes to the regular weather patterns and seasons on Earth, and more extreme weather, such as heat waves and storms.

A world without bees

If the worst-case scenario did happen, would we survive without bees? The short answer is yes! There are ways that we could make up for the loss of bees, but these solutions would be difficult, expensive, and not as good as the real deal.

Without bees to **pollinate** many of our **crops,** we'd have to eat crops that are pollinated in other ways, such as by the wind. Luckily, most grains, such as wheat, rice, and corn, do not rely on bees for pollination, so we wouldn't starve. However, since many fruits and vegetables are pollinated by bees and other insects, we'd miss out on lots of **nutrients** necessary for a balanced diet. Our meals would also be quite boring!

DID YOU KNOW?

Wheat can pollinate itself without the help of bees, other animals, or the wind.

We could also pollinate some previously bee-pollinated crops by hand or even with the help of robots! Scientists are developing different types of robots that could potentially help to pollinate crops if bee numbers continue to fall.

Hi! I'm Yu Gu. I'm a Chinese engineer who specializes in robotics. My team and I work on different robots that could help farmers pollinate their crops. One of our recent projects is StickBug, a six-armed robot that can pollinate flowers in a greenhouse. As well as using its arms to transport **pollen** from flower to flower, StickBug can also map out the greenhouse to discover which flowers need to be pollinated. We hope we never have to use robots like StickBug instead of bees, but it's good to be prepared, just in case!

THINK ABOUT IT!

What features would a robot need to have to allow it to pollinate a flower? Draw your own design.

A small paintbrush is a good tool for transporting pollen from flower to flower.

However, these methods of pollination would be much more expensive than natural pollination by bees. Farmers would have to hire huge numbers of workers to pollinate crops by hand or invest in expensive robot pollinators. They'd have to charge stores more for their produce so as not to lose money, and we'd end up paying much more for our groceries.

A WORLD WITHOUT BEES

It would be much harder to make up for the loss of bees in the natural world. Unlike with **crops**, humans wouldn't be able to replace bees by **pollinating** wild plants by hand or with robotic pollinators. Instead, we'd have to sit back and let nature find a new balance without bees or the flowering plants that they pollinate.

At first, many **ecosystems** would be hugely affected. We'd see many species of flowering plants become **endangered**. Certain plants that are almost only pollinated by bees, such as bee orchids, would probably become **extinct**. Other species that depend on bees for food, such as bee-eaters, would suffer greatly as well.

However, nature is adaptable. Over the history of our planet, there have been several major events that have made many animals extinct, such as the asteroid impact that led to the extinction of the dinosaurs. Some species have survived every one of these events and learned how to adjust to the new circumstances. So, it's very likely that the natural world would recover and adapt to the loss of bees.

The bee orchid flower looks like a female bee! Male bees climb inside the flower to try and mate with it and end up pollinating the flower without realizing it!

DID YOU KNOW?

After a huge asteroid impact, about 75 percent of life on land and 30 percent of ocean life, including dinosaurs, became extinct.

Without bees, other insects, such as wasps and butterflies, would take over the pollination of any surviving flowering plants. Wind- or self-pollinated plants would grow in the gaps left by the flowering plants that didn't survive. Animals that depend on bees for food would eat other flying insects, such as flies. The natural world would never be the same again, but most living things would survive.

Now I'm queen bee!

THINK ABOUT IT!

If bees disappeared, what changes would bee-eating animals have to make? How could they adapt?

Conclusion

Bees may be small, but they are mighty important! As we've seen, they help **pollinate** flowering plants and many of the **crops** we depend on for food, as well as providing food for other animals in many **ecosystems.** If bees disappeared, life on Earth would change forever. But sadly, despite their importance, bees are in serious danger.

DID YOU KNOW?

A quarter of all known bee species have not been seen since 1990.

Organic farms are my favorite!

However, it isn't too late to save bees from disappearing. People are responsible for most of the problems facing bees, such as **climate change, habitat** loss, and **pesticides.** Now that we know the impact that our behavior is having on bees, we must change it, so that we don't do them any more harm.

You can do your part to help bees by planting bee-friendly plants on windowsills or in your yard, or by switching to **organic** produce where possible. See if you can get involved with a **citizen science** project to help monitor bee **populations** in your local area. The information gathered in these projects helps bee scientists with their research.

Scientists can help to protect bees from other threats, such as diseases and pesticides. They have already made great progress with new developments, such as the first bee **vaccine!** Hopefully, we will soon see more effective treatments for other deadly bee diseases.

Making our planet more bee-friendly will do much more than just help bees. It will also help ensure the survival of many flowering plants, important crops, and other animals that depend on bees for food, such as bee-eaters.

However, if we don't take action now, it may be too late for bees. Let's hope, for the sake of the plants, our food supply, and the planet, that 100 years from now, our buzzing, striped friends aren't just a thing of the past.

Summary

So, what would happen exactly if bees disappeared? Check your understanding of the information in this book.

Bees' **habitats** are being destroyed. The land is built on or turned into farmland.

Bees are dying from diseases and **parasites.**

Climate change caused by human activity is affecting our seasons. Plants are flowering at different times, which means less food for bees.

Bees become **extinct.**

Many important **crops,** including fruits, vegetables, and nuts, aren't **pollinated** and don't grow properly.

We use artificial pollination (robots/by hand) to grow crops that were previously pollinated by bees. This will make them very expensive.

We end up mostly eating grains, which are pollinated by the wind or by the plants themselves. Our diet won't be very interesting or balanced, but we won't be hungry!

Bees are becoming very sick and dying because of **pesticides** used on fields of crops and other plants.

In some areas, **invasive species** are putting bees in danger.

Without these flowering plants, there is much less food for wild animals and fewer places for them to shelter.

Many wild plants aren't pollinated. This means that they can't produce seeds or grow into new plants.

Animals that eat bees go hungry.

Some plants and animals will become **endangered** or extinct, while others will adapt to survive in a world without bees.

THINK ABOUT IT!

Do you think it's likely that bees will become extinct in your lifetime? Why or why not?

Grandma, what were bees like?

Now I'm a butterfly-eater!

43

Glossary

agriculture—farming

carnivore—an animal that only eats other animals for food, like a lion—roar!

citizen science—scientific research done by ordinary people (just like you!) to help scientists with their work

climate change—changes in the world's weather, in particular, an increase in temperature, which scientists believe are mainly due to human activity

colony—a group of the same type of animals that live together in the same place

crop—a plant grown for food, such as apples, carrots, or potatoes

ecosystem—all of the living things in an area and the relationship between them

endangered—at risk of dying out because there are few of them left (oh, no!)

extinct—an extinct animal or plant no longer exists on Earth because its entire species has died out, just like the dinosaurs

food web—a diagram that shows all of the living things that depend on each other for food in an ecosystem (see page 23 for an example!)

fossil fuel—a fuel, such as natural gas, oil, or coal, which was formed over millions of years from the remains of animals and plants

global warming—an increase in temperatures on Earth due to the greenhouse effect

habitat—the place where an animal or plant usually lives

herbivore—an animal that only eats plants for food, such as a rabbit

hibernation—when an animal spends a period of time in deep sleep, usually during the winter (Zzzzzzz!)

immune system—cells and other parts of a body that protect it from disease

invasive species—a species that has arrived in a new area where it didn't previously live and is now causing harm to the other species that live there

larva (plural: larvae)—an early form of an insect, which has left its egg but not yet grown into its adult form

nectar—a sweet liquid found in flowers

nutrient—something that living things need in order to grow

omnivore—an animal that eats plants and other animals for food, such as a bear

organic—grown or raised without the use of pesticides or fertilizers

parasite—an animal or plant that lives on another type of animal or plant and feeds from it

pesticide—chemicals used to kill insects or other animals that harm crops

pollen—a powder found in flowers used for reproduction

pollinate—to transfer pollen from one flower to another for reproduction

population—how many animals or plants of the same type live in an area

predator—an animal that kills and eats other animals for food – watch out!

prey—an animal that is killed and eaten by other animals

reproduce—to produce new, young animals or plants

vaccine—a substance put into a living thing's body which protects them from disease by making them produce antibodies

virus—a very small living thing that causes disease

Review and reflect

COMPREHENSION QUESTIONS

Bees in crisis
- How has habitat loss due to human building and farming affected bees?
- What is an invasive species? Which invasive species is threatening bees?

Time to act!
- What can you do to help protect bees and keep them from dying out?
- Who is Marla Spivak and what is she studying?

Perfect pollinators
- Based on Dr. Jiandong An's findings, why should farmers support bees as pollinators?
- In your opinion, why should we be concerned that bees might not be around to help plants reproduce?

A world without bees
- How would your life change if the bees died out?
- Earth has experienced other major events that have led to extinctions in the past. With that in mind, what is likely to happen to Earth in the long term if the bees died out?

A balanced ecosystem
- How does looking at the food web on page 23 help you think about what might happen to other animals if all the bees died out?
- The author explained that ecosystems are carefully balanced. What might this mean?

Conclusion and summary
- After reading this book and considering what would happen if the bees died out, what is your biggest takeaway? Why?

MAKE A CHAIN OF EVENTS!

Creating a paper chain can help you explore and visualize how cause and effect relationships can be thought of as a sequence of events.

You'll need:
- Pencil
- Scratch paper
- Pens or markers
- Stapler and staples
- Strips of paper (2 colors, if possible)

Instructions:

1. **Select a focus:** Choose a specific aspect from the book that caught your attention—it could be an event, a factor affecting bees, or a consequence of their decline.

2. **Brainstorm causes and effects:** On a sheet of scratch paper, brainstorm and list the causes and effects related to your chosen focus. Think critically about the factors that contributed to or resulted from your focus. You can always look back in the text for ideas!

3. **Write on strips:** Write each cause and each effect on its own strip of paper. If you have different colored paper, use one color for the cause strips and the other for the effect strips.

4. **Create the paper chain:** Organize your strips into causes and effects. Start forming a paper chain to show how a cause leads to an effect. Use the stapler to connect the two strips. Continue adding cause and effect strips as links in your chain. When you've finished, you should be able to start at the beginning of your chain and read through each chain link in a logical order.

5. **Linking multiple chains:** If your focus has multiple causes or effects, you can create additional chains and link them together to show how complex cause and effect relationships can be!

Write about it!

Look at the paper chain you created and how the causes link to effects (which in turn link to other causes!). How might breaking a link in the chain impact the overall sequence of events?

World Book, Inc.
180 North LaSalle Street
Suite 900
Chicago, Illinois 60601
USA

For information about other World Book publications, visit our website at www.worldbook.com or call 1-800-WORLDBK (967-5325).

For information about sales to schools and libraries, call 1-800-975-3250 (United States), or 1-800-837-5365 (Canada).

© 2024 (print and e-book) by World Book, Inc. All rights reserved. No part of this publication may be reproduced, stored in a retrieval system, or transmitted in any form or by any means (electronic, mechanical, photocopying, recording, or otherwise) without written permission from World Book, Inc.

WORLD BOOK and the GLOBE DEVICE are registered trademarks or trademarks of World Book, Inc.

Library of Congress Cataloging-in-Publication Data for this volume has been applied for.

What Would Happen If...?
978-0-7166-5448-3 (set, hc.)

All the Bees Died Out?
ISBN: 978-0-7166-5449-0 (hc.)

Also available as:

ISBN: 978-0-7166-5455-1 (e-book)
ISBN: 978-0-7166-5461-2 (soft cover)

Staff

Editorial

Vice President
Tom Evans

Editorial Project Coordinator
Kaile Kilner

Curriculum Designer
Caroline Davidson

Proofreader
Nathalie Strassheim

Graphics and Design

Senior Visual
Communications Designer
Melanie Bender

Digital Asset Specialist
Rosalia Bledsoe

Written by Izzi Howell
Illustrated by Paula Bossio

Developed with World Book by
White-Thomson Publishing LTD
www.wtpub.co.uk

Acknowledgments

4-5	© 9dream studio/Shutterstock; © irin-k/Shutterstock
6-7	© Maksim Safaniuk, Shutterstock; © Mirko Graul, Shutterstock; © frees/Shutterstock; © Sander Meertins Photography/Shutterstock
8-9	© slowmotiongli/Shutterstock; © Alexandr Zagibalov, Shutterstock
10-11	© kosolovskyy/Shutterstock; © Beekeepx/Shutterstock
12-13	© Maarten Zeehandelaar, Shutterstock; © rospoint/Shutterstock; © Ian Redding, Alamy Images; © aaltair/Shutterstock; © Serhii Hrebeniuk, Shutterstock; © alfotokunst/Shutterstock; © Daniel Ostroukhov, Shutterstock; © Protasov AN/Shutterstock
14-15	© Nature Picture Library/Alamy Images; © irin-k/Shutterstock; © Brais Seara, Shutterstock
16-17	© 5D2/Shutterstock; © Jamie Tsai, Shutterstock
18-19	© Ines Behrens-Kunkel, Shutterstock; © Dima Moroz, Shutterstock; © Ruslan Semichev, Shutterstock; © Denis Nata, Shutterstock; © Roman Samokhin, Shutterstock; © grey_and/Shutterstock; © Maks Narodenko, Shutterstock; © Flower Studio/Shutterstock; © Nataliia K, Shutterstock; © monticello/Shutterstock; © azure1/Shutterstock; © Hortimages/Shutterstock; © topseller/Shutterstock
22	© Teresa Jane D, Alamy Images
25	© Stock Media Seller/Shutterstock
26-27	© Jonas Listl, Shutterstock; © critterbiz/Shutterstock
28-29	© Daniel Beckemeier, Shutterstock; © vitasunny/Shutterstock
30-31	© Fotokostic/Shutterstock; © Jesse Franks, Shutterstock
32-33	© Diyana Dimitrova, Shutterstock; © aslysun/Shutterstock
34-35	© Philip Yb Studio/Shutterstock; © ssuaphotos/Shutterstock; © Eco Print/Shutterstock; © Tory chemistry/Shutterstock; © Aliaksei Marozau, Shutterstock
36-37	© Nitr/Shutterstock; © nnattalli/Shutterstock; © grey_and/Shutterstock
38-39	© Stephan Morris, Shutterstock; © Philip Bird LRPS CPAGB/Shutterstock
40-41	© Firn/Shutterstock; © Stanimir G.Stoev, Shutterstock
44-45	© conzorb/Shutterstock; © phichak/Shutterstock; © Sander Meertins Photography/Shutterstock
46-47	© Roi and Roi/Shutterstock; © Sander Meertins Photography/Shutterstock

www.ingramcontent.com/pod-product-compliance
Lightning Source LLC
Chambersburg PA
CBHW060943170426
43197CB00023B/2974